Antonio Lima-de-Faria
Science and Art are Based on
the Same Principles and Values

© 2020 Artena Anarchist Press, Artena, Italy

ISBN 978-8-894-05053-0

All rights reserved. No part of this book may be reproduced, stored in a retrieval system, or transmitted, in any form or by any means, electronic, mechanical, photocopying, recording, or otherwise, without the prior written permission of the publisher.

Artena Anarchist Press
Via Vittorio Emanuele, 35
00031 Artena (Rome), Italy
www.artenarchist.com
info@artenarchist.com
@artenarchist

Printed in November 2020

SCIENCE AND ART
ARE BASED ON THE SAME
PRINCIPLES AND VALUES

Antonio Lima-de-Faria

Professor Emeritus of Molecular Cytogenetics
Lund University, Lund, Sweden

Artena Anarchist Press

About the Author

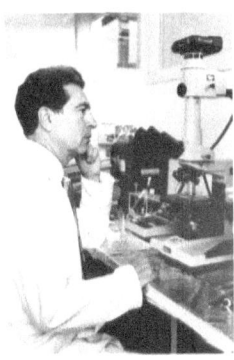

Antonio Lima-de-Faria is Professor Emeritus of Molecular Cytogenetics at Lund University, Lund, Sweden.

After editing the *Handbook of Molecular Cytology* 1969, Antonio Lima-de-Faria wrote *Molecular Evolution and Organization of the Chromosome* 1983, and *Evolution without Selection. Form and Function by Autoevolution* 1988. This last book was translated into Russian, Japanese and Italian. Later he wrote *Biological Periodicity. Its Molecular Mechanism and Evolutionary Implications* 1995 (translated into Japanese) and *One Hundred Years of Chromosome Research and what Remains to be Learned* 2003. Subsequently appeared *Praise of Chromosome "Folly." Confessions of an Untamed Molecular Structure* 2008 (translated into Russian), *Molecular Geometry of Body Pattern in Birds* 2012 and *Molecular Origins of Brain and Body Geometry* 2014. Recently appeared *Periodic Tables Unifying Living Organisms at the Molecular Level. The Predictive Power of the Law of Periodicity* 2017, which was praised, by American and English colleagues, as a "wonderful and pioneering work." His work in molecular biology has won him, among other honors, the decoration Knight of the Order of the North Star by the Swedish King and Great Official of the Order of Santiago by Portugal's President. He is member of five scientific academies.

The Author may be contacted at johanessenmoller@icloud.com

TO THE READER

From the beginning the intention was to treat this subject in book form, but it was soon reduced to an essay as a booklet. Finally it took the form of a pamphlet or of a manifesto.

You are listening to a lonely wolf howling in the immensity of the night sky.

Science is written, and art is painted, not with ink or colors, but with one's own blood. There is no place for the meek.

The following pages may appeal to some but not necessarily to others. As Louis Pasteur pointed out, "Only the prepared mind is able to see the phenomena that no one sees." The same seems to be the case for values and principles. You can only awake them in those who already carry their seeds in the inner of their minds.

CONTENTS

About the Author	v
TO THE READER	vii

PART 1

Where do We Stand Now	3
The Demand for a Future Full of Joy	5
Golden Age of European Culture — The 1800s	6
They Transformed the World for Centuries to Come	9
"The Spring Time of Peoples" — The 5 Simultaneous Revolutions that Shocked Europe in 1848	14
Einstein Lived Next Door to Leading Anarchists and Revolutionaries	16
Four European Generations	18

PART 2

How Music is Defined at Present	23
Musicians Defined Music	24
The Finest Music is Made Like a Shoemaker Makes a Shoe and the Same Holds for Science	26
The Relationship Between Music and Mathematics was Established as Early as the 5th Century B.C.	28
Bartók's Music was Based on the Golden Section	31
"Songbirds are Pythagoreans"	33

Singing in Birds, Whales and Humans has a Molecular Basis	35
The Songs of Birds Have Been Written Down as Music — Most of Their Genes are Similar to Those of Humans	38
Mozart Translated into Music the Song of a Starling and Used it in his Piano Concerto No. 17	41
The Music of Molecules — Alexander Borodin was at the Same Time a Leading Chemist and a Leading Composer	43
The Music of Proteins	45
Music Written from the Chemistry of DNA — "Construction of Coding Sequences is Based upon the Principle of Musical Composition"	47
In Music You Have to be Unorthodox as was Chopin	51
Mussorgsky Complained that "Whereas he Often Heard Painters or Writers Express Live Ideas, Musicians, to his Knowledge, Never Did Anything of the Kind"	53
Claude Debussy Remained All his Life a Big Child	54
Berlioz — The Creator of the Modern Large Orchestra	56
The Musician Giuseppe Verdi was Strongly Influenced by the Poet Victor Hugo	58
Music can be Converted into Graphic Art	59

PART 3

Definition of Art	63
Beauty and Subtlety Untainted by the Claims of Necessity	64

Everything is Beautiful for the Artist	65
Beauty is not a Quality of an Object and Few People Experience it in its Totality	67
A Musician's View of What is Beauty	70
A Biologist Asserts that "Beauty is a General Quality of Animate Nature"	71
For a Painter Beauty is Cosmic and Universal	75
Beauty is "A Manifestation of Secret Natural Laws"	77
Mathematical Creation is Synonymous with Beauty	78
For Plato Beauty is Absolute	80
The Conclusion is the Same, Irrespective of Whether You are a Scientist or an Artist	81

PART 4

Definition of Truth	85
How to Distinguish Truth from Fraud — Science is a Frightening Experience	86
"The Entire Approach Emphasizing 'Relative' Truth Seems to me a Piece of Humbug Masquerading as an Academic Discipline"	88
For Matisse there is an Inherent Truth — "Exactitude is Not Truth"	90
Truth is an Obligatory Quality of Creative Minds	92
What Looks Harmonious is Deformed. The Beauty of the Parthenon is a Contrived Illusion Created Intentionally by Greek Architects	94
Feminine Beauty was Also Intentionally Modified	97
Science is a Lie That Allows Us to Come Closer to the Truth	101

For Harold Pinter "Truth in Drama is Forever Elusive" but Real Truth is to be Defended	106

PART 5

Definition of Science	111
A Chemist's Reflections in the 20th Century	112
Pavlov's Testament — Science Demands a Whole Life	115
The Antithetic Nature of a Scientist's Qualities and Faults	119
It Seems that Scientists are Like Everybody Else	123
Occupation and Preoccupation	125
Writing on Horseback, in Trains or in Airplanes	127
A Classic Work was Written in a Week	129
The Fermentation of Ideas in Science Takes Many Years	131
In Art an Idea Also Takes a Long Time to Mature	134
Monumental Works	137

EPILOGUE

Geometry and the Order of Nature were the Main Preoccupation of Archimedes	145
"There is an Innate Principle of Justice and Virtue"	147
The Emperor Marcus Aurelius Has the Last Word	150
References	155

PART 1

WHERE DO WE STAND NOW

Science has become an industry controlled by the multinationals. Art is a source of economic investment for billionaires. We are witnessing, not only an inversion of values, but also a destruction of principles that so far have directed our most precious activities.

At present a wave of obscurantism is spreading over Western countries affecting both science and art in a deadly way. It becomes mandatory to stave off this movement by defining precisely the way science and art are created and to clarify the basic principles on which they are established. The scientific community has already reacted to this situation but the problem continues to be treated marginally. The reason may lie in the fact that modern technology has been most successful in transforming our daily lives and in allowing us to conquer outer space. These impressive achievements have, to a large extent, made us dumb, making it difficult to perceive the danger that lies ahead. Hence, there is a pressing need to bring forward the original sources in which, leading scientists and renowned artists, explained the principles that they followed in their discovery of novel phenomena and in the creation of unique works of art. It turns out that both types of minds speak the same language. There is a basic denominator that unites the human endeavor.

THE DEMAND FOR A FUTURE FULL OF JOY

Today's European Union is originally, and still remains, an economic union. It started in 1957 with the Rome Agreement, which united the economic interests of the large steel and coal industries of central Europe. It continued to grow by including new member countries, others were partly associated and finally some left the Union (Great Britain). It has gone through permanent economic and political convulsions mainly due to the fact that it lacks a common ideology. Its objective is "economic liberalism."

But like most social movements it has looked for some form of ideology and this is why it adopted the 9th Symphony of Beethoven as its hymn.

This is not accidental because Beethoven's 3rd and 9th symphonies were directly inspired by the French Revolution and the sense of joy that was emanating from a belief in a better future.

GOLDEN AGE OF EUROPEAN CULTURE — THE 1800s

The culture and values on which Western Culture stands today are mainly based on a unique period in the history of Europe.

This is an age that is comparable to the Renaissance that took place in Portugal, Spain, Italy and other European countries in the late 1400s and early 1500s. The outstanding achievement of scientists and artists that emerged at that time had not existed for over 1000 years and could not be repeated. This was the time of <u>Vasco da Gama</u>, who discovered the way to India, of <u>Fernando Magellan</u> who sailed around the globe and of <u>Christopher Columbus</u>, who found the maritime way to the American continent.

The world discoveries were not made by adventurers but were based on a strict scientific basis as demonstrated by the diaries of each voyage which have been preserved. The list of artists is a long one from

Michelangelo, the sculptor and painter, to the artist Albrecht Dürer, and many others. Among the scientists are Leonardo da Vinci, with a series of new inventions, Copernicus who developed his heliocentric system of the planets, Gerardus Mercator who published a map of the world based on a new projection, and Tycho Brahe who observed supernova. The sources of this sudden eruption in science and art were due mainly to the economic expansion resulting from the discoveries started in the early 1400s.

It is also the economic explosion caused by colonization and industrialism that led again to this second golden age of European culture.

The previous century had seen the British engineer James Watt (1736-1819) invent the modern steam engine. Machines would replace manpower for centuries to come.

But it was the birth of the German chemist Friedrich Wöhler in 1800 (1800-1882) which marked the beginning of the new century. Until then organic chemical compounds were thought to be different from inorganic ones carrying a "vital force" essential for the chemistry of life. In 1828, <u>Wöhler</u> disposed of this idea with a simple experiment. He synthetized urea, a typical animal compound from inorganic chemicals. This was a deadly blow to obscurantism.

THEY TRANSFORMED THE WORLD FOR CENTURIES TO COME

The list that appears below starts with Jules Verne the French novelist, who recognized that the advances in chemistry, physics and other areas would bring humanity to new boundaries. In his vision he invented submarines, airplanes and rockets to reach the moon, Jules Verne was called the "underground revolutionary."

It is remarkable that the leading participants in this golden age are all born within the period 1800-1899. They changed the face of Europe and of the World in every respect.

The individuals who participated actively in this cultural transformation are named below with country and their year of birth. It has been difficult to limit the list, and that obviously implies injustices.

Jules Verne, author, visionary of future technological development, France, 1828.

Friedrich Wöhler, synthesis of urea, disposes of "vital force," Germany, 1800.

Konrad Röntgen, discovery of X-rays, Germany, 1845.

Alexander Bell, inventor of telephone, Scotland, 1847.

Guglielmo Marconi, Marquis, wireless telegraphy, Italy, 1874.

Leo Tolstoy, Count, author and social reformer, Russia, 1828.

Mikhail Bakunin, anarchist, Russia, 1814.

Peter Kropotkin, Prince, revolutionary, anarchist, Russia, 1842.

Karl Marx, socialist, Germany, 1818.

Vladimir Lenin, communist leader, Russia, 1870.

Tse-Tung Mao, communist leader, China 1893.

Bertrand Russel, philosopher, England, 1872.

Gregor Mendel, rules of heredity, Austria, 1822.

Charles Darwin, expansion of Lamarck's concept of evolution, England, 1809.

Claude Bernard, pioneer of experimental medicine, France, 1813.

Ivan Pavlov, conditioned reflex, Russia, 1849.

Santiago Ramon y Cajal, nervous system, Spain, 1852.

Rudolf Virchow, pathologist, Germany, 1821.

Louis Pasteur, chemist and biologist, France, 1822.

Sigmund Freud, founder of psychoanalysis, Austria, 1856.

Dmitri Mendeleev, periodicity of chemical elements, Russia, 1834.

Pierre Curie, radioactivity, France, 1859.

Marie Curie, radioactivity, Poland, 1867.

Bernhard Riemann, non-Euclidian geometry, Germany, 1826.

Henri Poincaré, mathematician, France, 1854.

William Thomson, Lord Kelvin, physicist, England, 1824.

James Clerk Maxwell, electromagnetism, Scotland, 1831.

Albert Einstein, relativity theory, Germany, 1877.

Niels Bohr, atomic nuclei theory, Denmark, 1885.

Max Planck, quantum theory, Germany, 1858.

Heinrich Hertz, physicist, Germany, 1857.

Joseph John Thomson, Sir, discovery of electron, England, 1856.

Ernest Rutherford, Baron, physicist, New Zealand, 1871.

Pablo Picasso, painter, cubism, Spain, 1881.

Édouard Manet, painter, Impressionism, France, 1832.

Auguste Rodin, Sculptor, France, 1840.

Hector Berlioz, music composer, France, 1803.

Frédéric Chopin, music composer, Poland, 1810.

Richard Wagner, music composer, Germany, 1843.

Giuseppe Verdi, music composer, Italy, 1843.

Claude Debussy, music composer, France, 1862.

Pyotr Ilyich Tchaikovsky, music composer, Russia, 1840.

The towering figures of this period were Einstein (physics) Mendeleev (chemistry), Pasteur (biology and medicine), Tolstoy (literature), Picasso (painting) and Verdi (music). They stand out from the others due to their deep humanistic engagement. In them, the scientist and artist, were allied with a strong sense of the dignity of humankind.

"THE SPRING TIME OF PEOPLES" — THE 5 SIMULTANEOUS REVOLUTIONS THAT SHOCKED EUROPE IN 1848

The silence of the laboratories, where the discoveries were made, was broken by tumultuous events.

Just in the middle of the century, the people of Europe demanded a novel society free from serfdom and industrial slavery.

The revolution occurred simultaneously in Paris, Prague, Vienna, Berlin and Milan.

These revolutions were short lived due to the intervention of large armies. Most revolutionaries died in the barricades. In Vienna the price was 4,000 lives, in Paris 3,000 were slaughtered. One scientist who defended the rights of the unprivileged was the pathologist Rudolf Virchow, who also fought in the streets of Berlin. He was subsequently dismissed from his university position (Hobsbawm 1998).

But there were some irreversible changes.

Serfdom was abolished in Germany in 1848, and in Russia and Romania in 1860.

Remarkable is that some of the main ideologists of the new social order did not emerge from the proletariat, but from the highest nobility.

One should also not forget that Karl Marx was the son of a jurist and married an aristocratic lady, Jenny von Westphalen.

Kropotkin (1962) was Prince Peter Kropotkin. As a young man, he was a member of the inner circle of the Czar of Russia. Later he developed his revolutionary views that cost him to be exiled in Siberia.

Tolstoy belonged to the highest nobility, his mother was Princess Mariya Volkonskaya and he became Count Leo Tolstoy (Greenwood 1975).

Both Kropotkin and Tolstoy experienced at close site the great injustice that pervaded their society. Both were behind the future revolution of 1917. The result was that Tolstoy was excommunicated by the Orthodox Church and he never received a Nobel Prize (Rolland 1928).

EINSTEIN LIVED NEXT DOOR TO LEADING ANARCHISTS AND REVOLUTIONARIES

Science and art are not made in a vacuum. They are deeply influenced, not only by the technology available at the time, but also by the socio-economic situation and the ideology that dominates. Einstein could not have developed his space-time theory in a medieval cloister where obeyance to dogma was compulsory, or even at any other period, when intellectual activity was partly controlled by the Church which led to the imprisonment of Galileo and the exile of Descartes.

It may sound surprising to include political revolutionaries together with scientists in the above list but without the new political and anarchist ideas of Marx and Bakunin, Albert Einstein may not have dared to put forward his equally revolutionary relativity theory.

It is not by accident that Lenin, Kropotkin and Einstein lived in Switzerland at the same time, and that the three were refugees. Lenin was in Switzerland and mainly in Geneva during 1901-1908. Einstein lived in Zurich where he became a Swiss citizen. The relativity theory was published in 1905. Also living in Zurich, at that time was Kropotkin who came to the city in 1872.

International publications demanding the new organization of the working class and revolutionary journals were the order of the day.

FOUR EUROPEAN GENERATIONS

One is not usually aware that the values that your children and grandchildren espouse, and those that your parents defended, are essentially different. Hence, the language that you use may be quite foreign to them and what you try to explain may simply not be understood at all. The historical, political, technological and cultural landscape has changed radically during the latest four generations of Europeans. It is summarized in a nutshell.

(1) <u>Those who were born in the 1880s had a belief in an unlimited progress, both social and technological, that would change radically the structure of society.</u> They started to use trains, airplanes, televisions, telephones and the automobile. They also suffered the horrors of World War I.

(2) <u>Those who saw the world in the 1920s were confronted with a period of grave intellectual repression.</u> This took the form of fascism and nazism in a large number of

European countries. They also experienced the Spanish Civil War and World War II with over 40 million dead.

(3) The generation born in the 1960s were the children of the atom bomb. The world was on the brink of alienation. Between 1945 and 1963 the main military powers exploded atom bombs that encircled the globe in a cloud of radioactive fallout. This led to a critical, if not cynical attitude, in face of political events and the power of governments.

(4) The 2000s saw the appearance of the digital world. Reality and fiction became difficult to separate. Copying and plagiarism turned out to be "natural" processes. Limits became blurred, not only in science and art, but in everyday life.

The result has been a period characterized by an ideological vacuum, a relativity of concepts leading to different forms of legalized fraud.

At the same time great opportunities of communication, of education and of movement, on a global scale, have improved the opportunities of the individual and created possibilities to develop his or her own capacity. The control of disease has improved remarkably and their source has become better defined.

PART 2

HOW MUSIC IS DEFINED AT PRESENT

"Music is the art concerned with combining vocal or instrumental sounds for beauty of form or emotional expression usually according to cultural standards of rhythm, melody, and in most Western music, harmony. Both the simple folk song and the complex electronic composition belong to the same activity, music."

"Music is distinguished from *noise* by its ordered organization but by the later half of the 20th century 'noise' itself and silence became elements in composition and random sounds were used by composers" (*Britannica Academic*, Encyclopaedia Britannica, 2018, 2020).

Note that the word "beauty" is included in the definition.

MUSICIANS DEFINED MUSIC

Kaufmann (1947) is precise in her definition: "Melody is by definition a succession of musical sounds which have been organized into some kind of coherent shape or pattern." "Actually it is made of small elements called motives." And she adds "The method of varying the duration of tones is not haphazard, but an orderly, mathematical process." Modern physics confirms that there is a physical difference between noise and music — a basic difference that we are not usually aware of. "The roar of subway trains are not to be confused with music, for their waves, if photographed, would look jagged as a streak of lightning," on the other hand, "when a musical sound is uttered, its waves, are fully regular and properly timed."

Stravinsky (1936), explains what is music: "The phenomenon of music is given to us with the sole purpose of establishing an order in things, including, and particularly, the coordination between *man* and *time.* To be put into practice, its indispensable and single requirement is construction. Construction once completed, this order has been attained, and there is nothing more to be said." "It is precisely this construction, this achieved order, which produces in us a unique emotion having nothing in common with our ordinary sensations and our responses to the impressions of daily life."

He then adds that "the sensation produced by music is identical with that evoked by contemplation of the interplay of architectural forms. Goethe thoroughly understood that when he called architecture petrified music."

THE FINEST MUSIC IS MADE LIKE A SHOEMAKER MAKES A SHOE AND THE SAME HOLDS FOR SCIENCE

When he was writing one of his works Stravinsky discovered the "rigorous discipline" that it demanded.

He then remembered that Tchaikovsky, who was regarded, above all, as a lyrical composer, had written in one of his letters: "Since I began to compose I have made it my object to be, in my craft, what the most illustrious masters were in theirs; that is to say, I wanted to be, like them, an artisan, just as a shoemaker is... (They) composed their immortal works exactly as a shoemaker makes shoes; that is to say, day in, day out, and for the most part to order." It could not be more explicit.

Significant is that the sculptor Auguste Rodin says the same: "Do not rely on inspiration. It does not exist." "Accomplish your task as honest artisans."

Ivan Pavlov, the Russian discoverer of the conditioned reflex states: "First of all

— consistency. Of this very important condition of fruitful scientific work I can never speak without emotion. Consistency, consistency and again consistency. From the very beginning of your work train yourself to strict consistency in the acquirement of knowledge. Learn the *ABC* of science before you attempt to scale its peaks."

THE RELATIONSHIP BETWEEN MUSIC AND MATHEMATICS WAS ESTABLISHED AS EARLY AS THE 5th CENTURY B.C.

Pythagoras of Samos (c. 560-480 B.C.) was a Greek mathematician who emphasized that "the essence of all things is number." For him number was seen as a key to science.

He is remembered mainly from his work in geometry that led to the theorem on right-angled triangles named after him and treated in detail in Euclid's Geometry (c. 300 B.C.).

Pythagoras initiated the science of acoustics by being able to discover a relationship between the physical length of a vibrating string and the sound that it produced. The tones emanating from a stretched string were perceived as harmonious to the ear only when the lengths of string for the two tones have a simple number relation. For example, a length ratio of 2:1 corresponds to a musical octave. The way he found this relationship was solely experimental. On another occasion while passing the workshop of blacksmiths, he overheard the beating hammers. Examining the weight of the hammers he recognized that those sounding together the consonance of the octave were double in weight. Further he discovered how different weights led to mathematical ratios resulting in specific musical notes (Grout and Palisca 2001). A lyre, for instance, could not be used to play an acceptable melody when its strings were tuned at random. It was further established that there was a relationship between the measurable lengths of the chords of a lyre and audible harmony.

From these experiments it was inferred that musical relations correspond to intelligible mathematical principles, and since music is a paradigm of harmonious order, the Greeks sought to describe the ordering of nature in terms of similar numerical relations and principles (Hornblower and Spawforth 1999).

This way of thinking soon extended to scientific disciplines such as biology. Aristotle (384-322) followed this need to unify nature by systematizing in an orderly way the animal and plant world for the first time.

BARTÓK'S MUSIC WAS BASED ON THE GOLDEN SECTION

Music and mathematics are also related in other ways.

The Hungarian musician Bela Bartók (1881-1945) based his music on the Golden Section — a well-known mathematical rule. Leonardo Fibonacci was the Italian mathematician who introduced the Arabian numeral system to Europe in 1225, but he is mainly known for the "Fibonacci sequence" which consists of a series of numbers in which each successive number is the sum of the preceding two. This rule has many special properties and is evident in biological phenomena such as leaf and flower growth patterns, where it has been studied extensively by mathematicians, in collaboration with botanists (Jean and Barabé 1998).

In the case of plants, the angle of rotation about the axis of the plant between two consecutive flower units, expressed as a fraction of a turn, leads to a value which is called the golden section.

It is this extreme regularity, found in the organization of living organisms, that impressed Bartók, to such an extent, that he structured his music on the golden section. He used it to both create proportions between the musical movements and to the internal order of the composition. All the main intervals and repetitions, were distributed according to this rule. He actually intended to "write music according to the laws of nature" in nearly 2,000 published melodies.

"SONGBIRDS ARE PYTHAGOREANS"

Two thousand five hundred years ago the Greek mathematician Pythagoras was far from being aware that his studies not only elucidated the way the human mind behaves in science and music, but that his discoveries would apply also to the minds of animals, especially to those of songbirds.

In the book, "Nature's Music: The Science of Birdsong" (2004), the specialist on animal behavior Jeffrey Cynx reports in detail on the experiments carried out with starlings. The birds were trained to discriminate between melodies and to learn new ones. They seemed to memorize the true frequency of the song. He writes: "As to recognizing the ratios between notes, starlings, like humans, still respond to the 'missing fundamental' when the fundamental frequency of a harmonically structured sound is removed. This suggests that a bird's auditory system can fill in what is missing, based on Pythagorean ratios," and adds: "However, we can conclude that many songbirds are Pythagoreans and are able to hear frequency ratios, but also they perceive absolute pitch in a different way than we do."

SINGING IN BIRDS, WHALES AND HUMANS HAS A MOLECULAR BASIS

The molecular basis of singing and musical ability is best known in birds and humans. Like human youngsters, songbirds learn to vocalize by imitating the sounds of their elders. When zebra finches learn to sing, or adult canaries change their song, there is an increase in the protein produced by the *FoxP2* gene in the bird's brain. If the amount of this protein is reduced, by using virus-mediated RNA interference, there is a disruption in birdsong which becomes more variable and imprecise than that of controls (Haesler et al. 2007).

Other molecules such as hormones participate in the song process (Shen et al. 1995). Neural conversion of androgen to estrogen by the enzyme aromatase is an important step in the development and expression of masculine behavior in birds. The distribution of the aromatase messenger RNA has been mapped in the brain of finches and found to be widely expressed. The presence of estrogen receptors in the brain of several avian species has turned out to be unique to songbirds (Gahr et al. 1993).

Genetic aberrations of the human gene *FoxP2* impair speech production. The protein encoded by this gene is essential for full articulation of the human language. Mutations of *FoxP2* cause verbal disorders. This is actually the same gene that controls song production in birds and whose variation in activity in a bird's brain leads to song disorder (Haesler et al. 2007).

Humans are not the only mammals that sing and that can produce melodies. The humpback whale *Megaptera novaeangliae* has been extensively studied using deep-sea hydrophones combined with satellite tracking. These cetaceans migrate to tropical waters during the winter where they reproduce. Male mating strategy involves singing, as is the case in birds, as well as in humans under certain conditions. Lone whale males repeat long complex songs each lasting for about 10 min in an unbroken series. They can sing incessantly for over 24 h. They follow a hierarchical model in which "each song consists of a series of themes repeated in a specific order, the themes in turn are made up of phrases repeated a variable number of times" (Macdonald 2002). Songs from different years are quite distinct from one another. Besides, the songs of different humpback populations are independent. They can even adopt a new song from a different whale locality. Macdonald (2002) concludes that the humpbacks "learn the detailed acoustic structure of their songs." The parallel with human musical ability is striking.

THE SONGS OF BIRDS HAVE BEEN WRITTEN DOWN AS MUSIC — MOST OF THEIR GENES ARE SIMILAR TO THOSE OF HUMANS

Birds sing, whereas most other animals make a noise.

The sound emitted by a cow, a squirrel or by an elephant is not received by our brains as a song, *i.e.* a harmonious event. At best the sound emitted by an elephant can be compared to that of a trumpet, but not more than that. Other birds, like an owl, emit also sounds that are similar to single notes produced by an instrument.

But the sounds emitted by a nightingale, or a blackbird, are understood by our brain as a song. We even call each separate sound a note and the whole sequence is considered fully melodic.

These songs are in fact music, since they can be written down as notes. The composers <u>W.A. Mozart</u> and <u>Olivier Messiaen</u> made musical transcriptions of starlings' and chaffinches' songs respectively.

It is difficult to think that the nightingale has no sense of mathematics when it is able to sing: (1) over 10 different sounds (2) which range from a high pitch to the low notes of a cello and (3) that it produces them in a sequence which our brain recognizes as equally melodic as the music that we produce by our singing or by playing on a violin.

Different nightingales, even occupying nearby territories, sing differently. The main notes are the same, but there are evident differences in the melody. The presence of dialects in birds has been demonstrated by graphic records of their songs.

Before animals were believed to have different genes. A bird was supposed to have hardly any genes in common with a human. Following the DNA sequencing of the genome of the chicken (*Gallus gallus*) it became evident that a bird has many genes which are found in humans.

The gene number in *Homo sapiens* is 21,787 and in *Gallus gallus* is 17,709 (International Chicken Genome Sequencing Consortium 2004). About 60% of the chicken protein coding genes have a homologous gene sequence that has conserved the human gene order and orientation, revealing a high gene homology between humans and chickens (Lima-de-Faria 2012).

MOZART TRANSLATED INTO MUSIC THE SONG OF A STARLING AND USED IT IN HIS PIANO CONCERTO No. 17

M.J. West and A.P. King wrote an article in 1998 in the American Scientist with the title "Mozart's Starling." There, they describe how Wolfgang Amadeus Mozart purchased a starling in 1784. The bird and the composer had an intimate relationship, both were fascinated by each other. Mozart tried to imitate the starling's song transcribing it into music, which was used in his Piano Concerto No. 17.

The vocal capacity of starlings and their copying ability of the sounds of other animals, is known since Pliny (Gaius Plinius A.D. 23-79). He reported that starlings could mimic Greek and Latin, repeating words and sentences. The bird tried also to imitate Mozart since it had learned the themes from Mozart's scores. This mutual admiration lasted only three years, the bird dying suddenly.

Mozart was so taken by this loss that he composed a poem. He recited it at the grave of the starling, the funeral being preceded by a procession in which accompanying friends sang hymns. Such a display of affection may appear strange to outsiders. But you cannot be a composer, of Mozart's stature, without at the same time, being highly sensitive and respectful of other forms of life. For the composer, the bird was someone, who like him, tried to convey a musical message to the outer world.

THE MUSIC OF MOLECULES — ALEXANDER BORODIN WAS AT THE SAME TIME A LEADING CHEMIST AND A LEADING COMPOSER

The Russian musician <u>Alexander Borodin</u> (1833-1887) participated in the First Chemical Congress which was held in Karlsruhe, Germany in 1860. He belonged to the Medical Academy in Saint Petersburg where he was Professor of organic chemistry and a chemist of international repute.

He had combined his interest in music with chemistry since adolescence and it continued throughout all his life. As a chemist he is known for the pioneering discovery of two reactions which are at present part of the teaching in all laboratories.

As he recalled it was only when he was partly ill, and could not work in science, that he got the opportunity to compose his music. He is well known for the opera "Prince Igor" and several symphonies (Salter 1978).

THE MUSIC OF PROTEINS

The biologist Mary Anne Clark (Dunn and Clark 1999), describes in detail her search for "Life Music" or "Genetic Music." "I think that if somehow I could walk into a living cell, I would hear something similar — the ribosomes ticking away at the synthesis of proteins, playing out their amino acid sequences, note by note, according to a genetic score."

And she adds: "I was struck by the parallels between musical structure and the structure of proteins and the genes that encode them. Proteins also seem to be composed of phrases organized into themes."

As an example she chooses the amino acid sequence of *beta globin* which forms half of the hemoglobin molecule. Variations of the beta globin can be found in vertebrate species from all over the world. "Although these beta globin sequences are not identical in these species, they are similar enough that, if converted to music, they would be recognizable as variations of a common theme."

To concretize her idea she collaborated with the artist John Dunn. The sonification of protein data resulted in the production of an audio compact disc Life Music.

The merging of scientific knowledge with artistic expression produced music from the basic building blocks of life.

MUSIC WRITTEN
FROM THE CHEMISTRY OF DNA —
"CONSTRUCTION OF CODING
SEQUENCES IS BASED UPON THE
PRINCIPLE OF MUSICAL
COMPOSITION"

Susumo Ohno had a most original mind and was a leading chromosome researcher. He is known for his proposal that a main component of evolution was the process of gene duplication occurring in the chromosome. This idea was expounded in a book, "Evolution by Gene Duplication" (1970). Sex determination in mammals was also one of his areas of research.

In later years he became interested in the sequence of the bases along the DNA molecule, which are responsible for the molecular messages that constitute the genetic code. A DNA molecule consists of four bases abbreviated, A, T, G, C, associated with sugar and a phosphate backbone. The bases are grouped into well defined segments that delimit a structural gene or other types of DNA arrangements with diverse functions.

The DNA sequence in the chromosome is copied into RNA. This leads to the production of amino acids which finally assemble into proteins.

When this process was discovered the question arose. How many bases along the DNA molecule are responsible for the formation of a single amino acid? It took several years to decipher what became called the genetic code. It turned out that only three bases were necessary to code for one amino acid and that it was these triplets, called codons, found along the RNA which specified the 20 universal amino acids occurring in cell proteins.

Susumu Ohno was intrigued by these particular numbers and embarked on an analysis of the DNA molecule to find out what other particularities it could reveal. In a series of scientific articles he showed that, hidden inside the DNA sequences, there was a musical language. He actually was able to translate into musical scores the DNA sequences of several genes. In a work published together with the musician Marty Jahara they state: "Although most coding sequences of today are of considerable antiquity, they are nevertheless not unique. Instead each is composed of a number of recurring base oligomers [small segments of DNA] that are related to each other and their derivatives. Thus, they are constructed along the principle that governs musical composition."

This allowed these two authors to make a musical transformation of the 98-codon gene of the mouse, called: anti-NP^b IgV^H. Their article has the subtitle: "Construction of Coding Sequences is Based Upon the Principle of Musical Composition."

In another article, published by Ohno together with Midori Ohno they conclude: "Coding base sequences can be transformed into musical scores using one set rule. Conversely, musical scores can be transcribed to coding base sequences of long open reading frames." They went as far as to transcribe the initial portion of the Nocturne (opus 55 no. 1) for piano by <u>Frédéric Chopin</u>, into base sequences and their respective amino acids.

Conversely, they produced the musical transformation of the first 52 codons of the human X-linked phosphoglycerate kinase coding sequence. The piece was written for violin. And they added "If played on a violin this transformation is hauntingly melancholy, as though reflecting the Weltschmerz [Worldsorrow] of the gene that persevered for hundreds of millions of years."

One could have said that it was the voice of the painful birth of life that was being heard.

It is difficult to find a more pregnant example of the relationship between art and science, but still more significant is that classical music and the chemistry of the chromosome emerge as being based on a similar primordial pattern.

IN MUSIC YOU HAVE TO BE UNORTHODOX AS WAS CHOPIN

Frédéric Chopin (1810-1849) was a Polish child prodigy born in an aristocratic environment in which his father's intimates were poets, artists and scientists.

When he was an infant, the sound of music made him weep, and before he could write he began to compose. By the time he was eight years old his talents were discussed beyond the family circle (Huges 1943). Later, in Paris, after a period of desperate penury he moved in the highest circles among ambassadors, princes and ministers but he did not know how he got there and he added "I do not care for money, but only for friendship."

It was this friendship that he found in a remarkable woman, who, as a writer, used the pseudonym of George Sand. She had had many lovers, divorced her husband, and was now living on a large estate far from Paris. Another friend was also

Eugène Delacroix, the leading French painter during the Napoleon III period.

This was a remarkable and unique trio. In one room of the mansion Delacroix painted, in another Chopin composed and in a third part of the house, George Sand wrote her novels. She was so famous at that time, that the novelist Ivan Turgenev came from Russia, and traveled all the way from Paris to the estate only to meet her.

The Sonatas and the Ballades are among the best known pieces of Chopin. They baffled his contemporaries by being unorthodox and were considered formless. Today, they stand as "the most astonishing pieces of impressionist writing in the whole literature of the piano."

In Chopin's music the intervals of silence are as important as the notes that follow them. His melody is full of sorrow and pain. It is a message of the fight against his approaching death by tuberculosis (Hughes 1943).

MUSSORGSKY COMPLAINED THAT "WHEREAS HE OFTEN HEARD PAINTERS OR WRITERS EXPRESS LIVE IDEAS, MUSICIANS, TO HIS KNOWLEDGE, NEVER DID ANYTHING OF THE KIND"

Among the long line of Russian musicians, Modest Mussorgsky (1839-1881) stands as a main figure. He died early as a result of a bohemian life.

He was looking for "new ideas and new ideals," which obliged him to seek the company of writers and scientists who contributed to the maturing of his mind. His father was an officer of the Imperial Guards and he became a cadet in the Saint Petersburg regiment.

A part of Mussorgsky's music was left unfinished, and it was Tchaikovsky who later orchestrated it. Only in 1928 was the full score of his opera Boris Godunov issued by the Russian State Publishing Department.

CLAUDE DEBUSSY REMAINED ALL HIS LIFE A BIG CHILD

During the long history of European music, one finds two musicians who convey a particular sensation when we listen to them. We feel transposed to the birth of life, billions of years ago.

The sounds are pristine, and the composition does not allow for any compromise. One listens to the primeval awakening of nature. Obviously, one is referring to Stravinsky's "Rite of Spring" and Debussy's "L'après-midi d'un faune" (A Faun's afternoon) and "La Mèr" (The sea).

Claude Debussy (1862-1918) was a rebel. As a student in the Paris Conservatoire he was celebrated as the *"enfant terrible."* "I can only make my *own* music." "I will do my very best to satisfy a few people; as for the others, I don't care" (Abraham 1943).

He was very sensitive and attracted by everything subtle and delicate. He was fascinated by Mussorgsky as well as by oriental music. Not surprising he was influenced by the impressionist painters and symbolist poets, whose society he preferred to that of fellow-musicians.

His wife looked at him as a "big child." That was a very exact description of his mind. You need to have an unspoiled personality if you are to create something that is fundamentally original.

The French painter <u>Paul Cezanne</u> put it clearly: "One must see nature as no one has seen it before."

The creative mind is born anew every day.

BERLIOZ — THE CREATOR OF THE MODERN LARGE ORCHESTRA

It was the classic works of Greek and Roman literature, which Hector Berlioz (1803-1869) read in his early age, that influenced his music. They led the foundation of his love for the poet Virgil, who many years later led to his musical piece "Les Troyens." By 1841 he had started a tour of Europe that lasted for 9 years, spreading his fame as a composer. His "Symphonie Fantastique" is his major work. When it was performed in 1883, Paganini, the great Italian violinist and composer, was filled with praise for this most original music. Berlioz used in it a huge orchestra and introduced into classical music novel instruments, such as large church bells, to obtain the perfect dramatic content of the piece.

When he came to Berlin, the Emperor received him impressed by the music. "They tell me that you employ over 500 musicians." Berlioz replied: "Your majesty has been misinformed, it was only 435." In one of his later works he used 160 musicians, 3 soloists and a chorus of 98 for the vocal sections (Evans 1943).

Paganini, became his close friend and economically helped Berlioz to overcome his poverty.

THE MUSICIAN GIUSEPPE VERDI WAS STRONGLY INFLUENCED BY THE POET VICTOR HUGO

"La Forza del Destino" (The Force of Destiny) opens with one of the most impressive scores in music history. It cannot be forgotten, due to its originality and powerful message of the fate of mankind. Verdi and Wagner, were born in the same year, and are the greatest figures of the Italian and German Opera. But whereas Wagner heralded a Middle Age view of values based on the church, on war and personal power, Verdi instead was struck by the personal tragedies of his contemporaries and the inequalities of a society blinded by money.

Verdi used some of the novels of the French poet Victor Hugo, who embodied the romantic liberalism of the period, as the basis of his music. "Strength and sincerity were Verdi's most valuable attributes as an opera composer" (Toye 1943).

MUSIC CAN BE CONVERTED INTO GRAPHIC ART

Abstraction in painting is considered to have been influenced by the realization that both color and sound are produced by musical vibrations of varying wavelengths. This approach, which dominated at the end of the 1800s and beginning of the 1900s, was pioneered by artists such as Turner, Monet, Kandinsky and others.

In 1905, F. Bligh Bond materialized such musical vibrations in forms that have a geometrical pattern and display various types of symmetry converting music into graphic art (Gibson 2003).

PART 3

DEFINITION OF ART

The term art encompasses diverse media such as painting, sculpture, printmaking, drawing, decorative arts, photography and installation." "The various visual arts exist within a continuum that ranges from purely aesthetic purposes at one end to purely utilitarian purposes at the other."

Particularly in the 20th century a debate arose over the definition of art, when in 1917, the *DADA* artist Marcel Duchamp submitted a porcelain urinal entitled "Fountain" to a public exhibition in New York City. "By the turn of the 21st century, a variety of new media, further challenged traditional definitions of art" (Britannica Academic, Encyclopaedia Britannica 2016, 2020).

BEAUTY AND SUBTLETY UNTAINTED BY THE CLAIMS OF NECESSITY

The debate on what was to be defined as beauty already raged in Antiquity. Plutarch (46?-120? A.D.), who masterly described the lives of the main figures of Antiquity, let us know the opinion of the Greek scientist Archimedes (287-212 B.C.), who was a mathematician, physicist and engineer.

"Archimedes concentrated his ambition exclusively upon those speculations whose beauty and subtlety are untainted by the claims of necessity. These studies, he believed, are incomparably superior to any others, since here the grandeur and beauty of the subject matter vie for our admiration with the cogency and precision of the methods of proof" (Plutarch, published 1965). The scientist is referring to the mathematical relationship, that can be demonstrated, between geometric forms, such as the cylinder and the sphere.

EVERYTHING IS BEAUTIFUL FOR THE ARTIST

Auguste Rodin (1840-1917) was the French sculptor considered to be "one of the greatest and most influential European artists of his period" (Chilvers 2003). Some of his best known sculptures are "The Thinker" and "The Gates of Hell."

Rodin's message of what is art was actually called his testament by Paul Gsell who in 1946 published his interviews with Rodin under the title "Auguste Rodin, l'Art."

The following sentences are extracts from Rodin's "Testament" translated from the French original.

"Nature ought to be your goddess. Have in her an absolute trust. Be sure that she is never ugly and limit your ambition to be faithful to her." "Everything is beautiful to the artist, because in every being and in every thing, his penetrating vision discovers the 'character,' that is, the interior truth which results in the form. And this truth, is beauty itself." "Be patient. Do not rely on inspiration. It does not exist. The sole qualities of the artist are wisdom, attention, sincerity, decision. Accomplish your task as honest workers. Be profoundly, aggressively true. Never hesitate to express what you feel, even if you are in opposition to the prevailing ideas." "The great question is to be a man before being an artist."

And <u>Rodin</u> finishes "Art is above all a magnificent lesson in sincerity" — the same could equally be said of science.

BEAUTY IS NOT A QUALITY OF AN OBJECT AND FEW PEOPLE EXPERIENCE IT IN ITS TOTALITY

Beauty is usually defined as a quality which makes an object to give pleasure to the esthetic sense as by line, color, form, texture, proportion, rhythmic motion, or other features. The popular notion of beauty considers it a "quality" limited to "objects" having nothing to do with Nature.

Besides, people assert that beauty either is a concept difficult to define or is relative. This seems to be so because few people actually experience beauty in its totality.

All people sense beauty to some degree, but that is far from the unique beauty that is only experienced by few. These are the great artists, scientists, philosophers and poets who penetrated into the depth of natural phenomena. For them there are no "different tastes" or "undefined sensations." Once you experience the profound beauty of: a mathematical equation, a chemical reaction, a centenary tree, a prehistoric painting, or a classic music piece that you cannot forget; you know where beauty lies. You get an imprint that is stamped in your mind for the rest of your life.

An old-fashioned view is expressed, by one of the leading art critics of the last century, Herbert Read, who tried to explain "The Meaning of Art" in his work of 1950. The value of this book lies in his extension of the concept of beauty, and of art, to objects that before were considered to be ugly or were called "primitive artefacts." He represented a departure from the conservative view of art that had dominated until then. According to the early view, beauty was to be found mainly in the classical Ancient world of Greece and Rome and in the imperialistic period of Victoria Queen of Great Britain. An imposing totem from the natives of North Canada, or a decorated boat carved by Indonesians had been excluded from art since they were considered disgusting products of inferior people.

Read writes: "The concept of beauty is, indeed, of limited historical significance. It arose in ancient Greece and was the offspring of a particular philosophy of life. That philosophy was anthropomorphic in kind; it exalted all human values and saw in the gods nothing but man writ large."

What Read describes is the origin of the concept of beauty in objects and how it changed with time and location. He is still a prisoner of the general idea that beauty is solely restricted to objects. Nature continues to be ignored.

It is true that objects, such as sculptures, paintings or cathedrals, have a beauty in themselves, but this beauty is only understood as an extension of the one that is found in natural phenomena. Or better formulated, as a film director put it: "the beauty that I see in a tree is in me."

Actually, beauty can be defined, and with precision.

A MUSICIAN'S VIEW OF
WHAT IS BEAUTY

To most people, even to those who temporally dealt with science or art, beauty appeared difficult to define. However, those who had a profound experience of this process did not hesitate to define it in exact terms.

Igor Stravinsky (1882-1971), the Russian composer, who changed music radically by liberating it from its classical tradition, stated in his opera "The Rake's Progress" (with Fable libretto by W.H. Auden and Chester Kallman): "Mother Goose: What is the secret Nature knows? Tom Rakewell: What beauty is and where it grows."

Nature knows what beauty is because it is intrinsic to its construction.

A BIOLOGIST ASSERTS THAT "BEAUTY IS A GENERAL QUALITY OF ANIMATE NATURE"

In the book by J. Arthur Thomson "The System of Animate Nature" (1920), Chapter 8 is called "The fact of beauty."

Thomson was professor of Natural History at the University of Aberdeen, United Kingdom, and his work is a detailed description of biological phenomena.

He writes: "A Synoptic View of Nature Must Include the Fact of the Pervasiveness of Beauty" and he adds: "Beauty is a General Quality of Animate Nature." In a most lucid treatment of the subject he summarizes his point of view as follows: "In an inquiry into the significance of Animate Nature, there is no getting past *the fact of Beauty*. There are curiously few general affirmations that we can make about Nature; one is that Nature is in great part intelligible or rationalisable, and another is that Nature is in greater part beautiful."

"In an endeavour to indicate what contribution Natural Science has to make to our general view of the world, it is impossible to pass over the pervasiveness of beauty in the realm of organisms. Scientific investigation has disclosed it in the microscopically minute, in internal structure, in the well-concealed — everywhere. What concerns us in this study is the interesting fact that all natural, free-living, fully-formed, healthy living creatures, which we can contemplate without prejudice, are in their appropriate surroundings artistic harmonies — a joy to behold."

"This thesis may be objected to on various grounds — that beauty is wholly in our minds, that our likes and dislikes are wholly due to individual and racial nurture, that there is no agreement as to what is beautiful; but it seems possible to meet these objections. Another series of objections, however, consists of evidence that the realm of organisms is spotted with ugliness; and to meet these it is necessary to emphasise the saving-clauses of our thesis, that it does not apply to the domesticated and cultivated, the diseased or crippled, the unfinished, the parasitic, and the freakish."

"The elements that make up the impression we call visual beauty are arrangements and combinations of lines and colours, and a pre-condition of the beautiful is some quality of satisfactoryness in this pattern. In the case of animals, and somewhat apart, pleasing movements may be added to the presentation. But the big fact is that the stamp or halo of beauty is on every free individuality, and if the straight lines and the curves, the patterns, the colours, and the apportionment of the colours be expressions of normal vital processes, and so with rhythmic movements, it becomes easier to understand why they appeal in a pleasant way to wholesome sensoria with the requisite freedom of response." "The consistent discernment and enjoyment of the beautiful cannot be attained on any easier terms than consistent discernment and enjoyment of the True and the Good."

It is significant that Arthur Thomson states clearly that beauty is everywhere and by that he means that not only is found in any organism, but is present at every level of investigation, such as in the minute protozoa seen only with the microscope.

Thomson had in mind the work of Ernst Haeckel (1834-1919) who drew and painted, with the utmost accuracy, hundreds of microorganisms. These revealed their impressive complexity and symmetric structure that concealed an ordered organization. One of his works covers 100 magnificent plates which incorporate not only the simple protozoa but other minute animals and plants which populate the seas (Haeckel 1904).

FOR A PAINTER BEAUTY IS COSMIC AND UNIVERSAL

The Dutch painter Piet Mondrian (1872-1944), who was one of the most important figures in the development of abstract art, published in 1919 "Natural Reality and Abstract Reality." Mondrian highly valued internal rather than external things and abstract rather than natural ones. He wrote: "The truly modern artist is aware of abstraction in an emotion of beauty; he is conscious of the fact that the emotion of beauty is cosmic, universal. This conscious recognition has for its corollary an abstract plasticism, for man adheres only to what is universal."

He promoted a new kind of rigorously geometrical abstract painting limiting himself to straight lines and basic colors to create an art of great clarity and discipline. For Mondrian "art reflected the laws of the universe revealing immutable realities behind the ever-changing appearances of the world" (Chilvers 2003).

BEAUTY IS "A MANIFESTATION OF SECRET NATURAL LAWS"

The poet and novelist Johann Wolfgang von Goethe (1749-1832) in his "Maxims and Reflections" extended his thoughts to a long range of subjects. Beauty is not missed being defined as: "Beauty is a manifestation of secret natural laws which without this appearance would have remained eternally hidden from us."

At that time natural laws were known in physics, thanks mainly to the work of Galileo and Newton, but the laws directing chemical and biological phenomena awaited to be discovered. Goethe anticipated that they ought to be found but called them secret because, as late as the French Revolution (1789), these laws had not yet been formulated.

MATHEMATICAL CREATION IS SYNONYMOUS WITH BEAUTY

Henri Poincaré (1854-1912) was not only one of the most eminent physicists and mathematicians of France but he left also his mark on science by writing two works that became classics: "Science et Méthode" 1908 (1947) and "La Science et l'Hypothèse" 1906 (1943).

For Poincaré mathematical creation was synonymous with beauty: "Now, what are the mathematic entities to which we attribute this character of beauty and elegance, and which are capable of developing in us a sort of esthetic emotion? They are those whose elements are harmoniously disposed so that the mind without effort can embrace their totality while realizing the details. This harmony is at once a satisfaction of our esthetic needs and an aid to the mind, sustaining and guiding. And at the same time, in putting under our eyes a well-ordered whole, it makes us foresee a mathematical law. The only mathematical facts worthy of fixing our attention and capable of being useful are those which can teach us a mathematical law. So that we reach the following conclusion: The useful combinations are precisely the most beautiful, I mean those best able to charm this special sensibility that all mathematicians know, but of which the profane are so ignorant as often to be tempted to smile at it." Cited by Ghiselin (1952).

Poincaré emphasizes three basic aspects. (1) Beauty rests on a well-ordered whole. (2) It leads to the prediction of laws. (3) Mathematicians have this sensibility but the ignorant profane are far from understanding it.

FOR PLATO BEAUTY IS ABSOLUTE

Chilvers (2003) mentions Plato's opinion on beauty.

"The basic aesthetic premises of abstract art — that formal qualities can be thought of as existing independently of subject matter — existed long before the 20th century. Ultimately the idea can be traced back to Plato, who in his dialogue *Philebus* (c. 350 B.C.) puts the following words into Socrates' mouth: 'I do not mean by beauty of form such beauty as that of animals and pictures but understand me to mean straight lines and circles, and the plane or solid figures which are formed out of them by turning-lathes and rulers and measures of angles: for these I affirm to be not only relatively beautiful, like other things, but eternally and absolutely beautiful.'"

It is straight lines, circles and solid figures, it is all the order inherent to geometry, that is at the basis of the absolutely beautiful.

THE CONCLUSION IS THE SAME, IRRESPECTIVE OF WHETHER YOU ARE A SCIENTIST OR AN ARTIST

What is beautiful to someone is not necessarily beautiful to another. This notion, that beauty is relative, can be considered irrelevant.

Those who had the opportunity to be confronted with beauty in its pure form, as they entered into the depth of the unknown realm of Nature, do not doubt of its unique origin.

For a musician (Stravinsky), a painter (Mondrian), a biologist (Thomson), a poet (von Goethe), a mathematician (Poincaré), a physicist (Archimedes), a philosopher (Plato), a sculptor (Rodin) — for all of them — beauty is intrinsic and universal.

PART 4

DEFINITION OF TRUTH

Truth is "the property of sentences, assertions, thoughts, or proportions that are said, in ordinary discourse, to agree with the facts or state what is the case." And it is added: "a dedicated pursuit of truth characterizes the good scientist."

The definition is followed by a discussion of its philosophical meaning but that is outside the scope of this book (*Britannica Academic*, Encyclopaedia Britannica 2009, 2020).

HOW TO DISTINGUISH TRUTH FROM FRAUD — SCIENCE IS A FRIGHTENING EXPERIENCE

Richard Feynman (1918-1988) was one of the leaders of the team that worked on the Manhattan Project to develop the atomic bomb during the 1940s. He is the physicist who developed the theory of quantum electron dynamics which is considered one of the keys to understand the Universe. It actually describes the interactions involving light (photons) and charged particles (electrons).

Feynman described science in the following way:

"Science is a way to teach how something gets known, what is known, to what extent things are known (for nothing is known absolutely), how to handle doubt and uncertainty, what the rules of evidence are, what to think about things so that judgements can be made, how to distinguish truth from fraud, from show... in learning science you learn to handle by trial and error, to develop a spirit of invention and of free inquiry which is of tremendous value far beyond science. One learns to ask oneself: 'Is there a better way to do it?'" (Gribbin and Gribbin 2018).

This definition puts the emphasis on the spirit of free inquiry, on the distinction of truth from fraud, and on the difference between science and obscurantism.

In a later connection he added: "You are describing how science is done. I know, for I have had the same beautiful and frightening experience." Discovery into the deep realms of the unknown involves such an impressive beauty that one feels frightened when confronted with it.

"THE ENTIRE APPROACH EMPHASIZING 'RELATIVE' TRUTH SEEMS TO ME A PIECE OF HUMBUG MASQUERADING AS AN ACADEMIC DISCIPLINE"

It has not only become a fashion to discredit the contribution of European scientists, and artists but it has even been made into a scientific discipline. Examples are <u>Louis Pasteur</u> and <u>Pablo Picasso</u> but many others have been the target of this lucrative industry.

The attack is so devastating and so serious that <u>Max Perutz</u> (1914-2002) Nobel Laureate, and one of the founders of Molecular Biology, was obliged to use a whole chapter in his book (<u>Perutz</u> 2003) to state: "In *The Private Science of Louis Pasteur* <u>Gerald L. Geison</u> a historian of science, claims to have deconstructed <u>Pasteur</u>, and to have produced 'a fuller, deeper and quite different version of the currently dominant image of the great scientist.'" <u>Perutz</u> concludes: "The entire approach emphasizing 'relative' truth seems to me a piece of humbug masquerading as an academic discipline."

Pablo Picasso was the victim of most violent attacks during his life time, but now is recognized as the greatest artist of the century.

But Perutz thinks that: "In science, truth always wins" (Ferry 2007).

FOR MATISSE THERE IS AN INHERENT TRUTH — "EXACTITUDE IS NOT TRUTH"

Henry Matisse (1869-1954) was one of the leading painters of the 20th century and one of the theoreticians of art. Other painters, like Pablo Picasso, refused to comment on their art, but Matisse rejoiced in explaining it.

One of his main statements is: "Exactitude is not truth." And he explains it clearly when referring to his masterly pictures: "These drawings seem to me to sum up observations that I have been making for many years on the characteristics of a drawing, characteristics that do not depend on the exact copying of natural forms, nor on the patient assembling of exact details, but on the profound feeling of the artist before the objects which he has chosen, on which his attention is focused, and the spirit of which he has penetrated." "My convictions on these matters crystallized after I had verified the fact that, for example, in the leaves of a tree — of a fig tree particularly — the great difference of form that exists among them does not keep them from being united by a common quality. Fig leaves whatever fantastic shapes they assume, are always unmistakably fig leaves. I have made the same observation about other growing things: fruit, vegetables, etc." (Flam 1973).

"Thus there is an inherent truth which must be disengaged from the outward appearance of the object to be represented. This is the only truth that matters" (Chipp 1968).

Matisse puts it "black and white": Truth is dependent on a profound feeling of a common quality. The only truth that matters is that which is inherent, not the outward appearance.

TRUTH IS AN OBLIGATORY QUALITY OF CREATIVE MINDS

For Johann von Goethe, the concept of truth was his main preoccupation: "The first and last thing demanded of a genius is love of truth." "He who is and remains true to himself and to others has the most attractive quality of the greatest talents."

"What is exact about mathematics except exactitude? And this, is it not the result of an innate sense of truth?" "Truth is constructive; error is unproductive, it only constrains us."

"Truth, so it is said, is situated at the central point between two opposing views. Not at all!" "Truth is god-like: it is not immediately perceptible; we are obliged to guess it from its manifestations." "Wisdom is to be found only in truth" (Goethe published 1998).

The ideas of "genius" and of "great talents," which were in use in the time of Goethe (1749-1832), have been replaced by those of an eminent scientist or artist. The difference is that the genius was a figure of the romantic age whereas the scientist and artist, are more close to the reality of today's technological society. However, what is central is that <u>Goethe</u> considers it an obligatory quality of the creative mind.

Besides, he points out that truth is an innate sense and denies its relativity, heralded by those who are outside its deep experience.

WHAT LOOKS HARMONIOUS IS DEFORMED. THE BEAUTY OF THE PARTHENON IS A CONTRIVED ILLUSION CREATED INTENTIONALLY BY GREEK ARCHITECTS

The position of Matisse on art is confirmed by that of the architects in Ancient Greece.

Honour and Fleming (2002) describe their activity in the utmost detail.

The Parthenon is considered the supreme example of the Doric Temple, a type of building which evolved in the course of the preceding two centuries. This architecttonical gem has become familiar by late imitations built throughout the Western World. However, its original purpose and peculiarities are too often overlooked.

Greek temples were not designed for ritual. Religious ritual was focused on the open-air altar where sacrifices were made to the gods. The temple was built to enshrine the statue of the deity to whom it was dedicated.

The lay-out of these sanctuaries reveals that the oblique view was that envisaged by the architect. This view appears to be perfectly rectilinear and regular. But this is a carefully contrived illusion. The lines are not straight nor are the columns equally spaced.

Greek architects introduced what were called "optical refinements" to compensate for what might have been disturbing visual effects. "The optical refinements of the Parthenon are so effective that they pass unnoticed until they are pointed out. The whole platform, for example, is very gently curved down from the center. The columns all slope inwards and bulge out to the top. They also appear to be spaced regularly, but the three of each corner are closer together than the rest."

These optical refinements are by no means peculiar to the Parthenon; they were employed, with variations, in all Greek temples of the fifth century B.C.

Thus, Matisse's dictum "exactitude is

not truth" prevailed in the Greek Temples. To convey the sense of order and harmony, which is at the basis of truth, the Parthenon had to be distorted, in a very specific way, which took several centuries to discover. It contains a message that is perceived by all mankind.

FEMININE BEAUTY WAS ALSO INTENTIONALLY MODIFIED

Ancient Babylonians, Persians, Egyptians and Greeks were after what was fundamental in the organization of nature. Their buildings are a concretization of their pursuit.

From the beginning the Egyptians had built temples with huge columns on their inside, which were symbols of permanence and harmony. The Greeks changed their position radically, they put the columns on the outside of the building to convey directly the harmony. Additionally, the Egyptians did not build their monuments as flat structures, but as ideal geometric figures. The pyramids were, from the start, cut at the top and built stepwise because the slope of these truncated pyramids was easier to calculate mathematically (Katz 1993). Later, as their scientific knowledge increased, the pure pyramid became the standard type.

The Greeks started by copying the

early human figures of the Egyptians, with a rigid body, placed flat against a wall and in profile. Soon this representation was abandoned to create three dimensional sculptures. But their female and male figures were now going to represent something different. Their sex was obvious, and well delimited, but the message had to go beyond that property.

The Greek ideal demanded that sculptors ought to transcend everyday appearances by selecting only the best models and eliminating all apparent flaws.

Plato was one of the theoreticians of the period. For him "all perceptible objects are imperfect copies, approximating to imperceptible ideas." Socrates, his contemporary, remarked that "you combine the best features of every one of a number of models and so convey the appearance of entirely beautiful bodies."

Later the Roman author Cicero (B.C. 106-44) told how, the painter Zeuxis, four centuries earlier, had employed five different women as models for a single picture of Helen of Troy "for nature has not refined to perfection any single object in all its parts." As Honour and Fleming (2002) point out Greek sculptors followed a set of rules, *i.e.* a *canon*:

1) First the human body had to be naked not for erotic reasons but to represent the proportions of its different parts, which, they had found, concealed exact relationships.

2) The muscles were well represented in male figures to express an ideal of indomitable force and power.

3) The head of a woman was not to represent any specific person, but was to be created by assembling: the long neck from one, the large eyes from another, and the straight nose from a third. This was a woman who never existed in reality, but who conveyed an ideal of serenity.

The Venus of Milo, in the Louvre Museum, Paris, is the iconic example of this type of sculpture. It represents a later period in Greek sculpture, because she is not anymore fully naked, but is covered by light and partly transparent clothes. Her diaper, and the turning of her torso, introduced a dynamic component that was a novelty.

To note is that the breasts of the Venus of Milo are very well formed in appearance, but they are not made to release a sexual attraction. They are rigid and solid structures being part of a total body harmony.

<u>Auguste Rodin</u>, understood this message well. He called the Venus of Milo a "gendarme," that means in French, a policeman. He was well aware of the fact that all her muscles had been made nearly invisible. She was stiff as a Greek column and did not exhibit life's irregularities and interior force.

She was indeed a wonderful conveyer of human dignity that the Greeks tried to concretize.

SCIENCE IS A LIE THAT ALLOWS US TO COME CLOSER TO THE TRUTH

Pablo Picasso was a Spanish painter who lived most of his life in exile in France. Since his art was so startling and since he changed his style so often, people asked him to comment on his paintings, something that he refused. Besides he seldom gave interviews or wrote about art. In one of the few occasions, he provided what is perhaps one of the best definitions of art: "Art is a lie that allows us to approach to the truth" (Chipp 1968). Art is always a deformation of the reality that we are confronted with. Otherwise it would not add anything to our view of the world. Art is one of the ways mankind uses to penetrate the hidden secrets of nature, it only uses a technique which is different from that of science, but the objective is the same. One expresses in colors, patterns, forms or musical sequences, hidden harmonies and relationships that the eye or the ear usually are not able to discern and convey. Picasso's definition of art could equally be applied to science. "Science is a lie that allows us to come closer to the truth." Science never conveys reality but only an interpretation of it. This interpretation depends primarily on the technology that is available at a given time. Instead of the "lie" being crystallized in colors or sounds, it is crystallized in mathematical equations. These establish relationships between phenomena that before could not have been discerned or unified. The mathematical equations are used in the most advanced sciences, such as theoretical physics and chemistry. In less advanced sciences like biology, psychology or sociology, equations are yet to be found. The relationships between the phenomena are so far expressed in more crude forms such as "rules." But the fundamental process involved is the same.

Even among scientists one finds the idea that science conveys "truths" or "final judgements." Such situations are foreign to science, they are the domain of religion which is built on dogmas. Science is assembled by putting together approximate interpretations that change as knowledge becomes deeper. Even the most exact laws in physics, such as those of gravity contain exceptions. And what are the exceptions? They are the tangible examples of the limitations of every scientific interpretation. The law — and the equations that define it — come close to a description of the actual reality. But — and that is the central point — they never represent a total and terminal explanation. They are, at best, a close approximation to reality.

According to Kepler's first law the motion of the planets, including the Earth, go around the Sun on an elliptical orbit, but the Sun is not at the center of an ellipse, it is at an off-center position. Also, the eccentricity of some planetary orbits, such as that of Mercury, is not negligible (Weinberg 2015).

However, it is to be noted that previous concepts are not discarded in their entirety but superseded, since they are based on knowledge that was carefully assembled. It is not the data available that are rejected but it is only their interpretation that is modified, simplified or enlarged. But one "lie" only substitutes another. The great value of science is that the "lie" is constructed in such a way that it allows us to predict better and better the phenomena that occur around us or in our own body. As such, it gives a power that no other human activity can convey. This is why we tend erroneously to speak of scientific truths.

Art also gives us a power, but of a different nature. It furnishes us with the great strength of enduring the vagaries of everyday life and the deformations that we are subjected to by a complex society. This has become inhuman in many aspects due to its technological expansion, and art allows us to cope better with it.

But more than that, real art, like science, leads into the depths of the unknown, and puts its finger mark on the virgin forests of our mind.

Art establishes relationships that we had not perceived before, and in this way allows us to better understand, and in consequence to better master, the universe that surrounds us. Rembrandt's self-portraits, Goya's The Disasters of War and Kandinsky's abstract triangles, furnish key insights into the origins of psychological problems and of social conflicts. In the tapestries of the French artist Jean Lurcat (1892-1966) the unity of life is the central theme.

FOR HAROLD PINTER "TRUTH IN DRAMA IS FOREVER ELUSIVE" BUT REAL TRUTH IS TO BE DEFENDED

In his Nobel Lecture "Art, Truth and Politics," on December 7, 2005, Harold Pinter included his statements on truth.

"In 1958 I wrote the following: Truth in drama is forever elusive. You never quite find it but the search for it is compulsive. The search is clearly what drives the endeavour. The search is your task. Sometimes you feel you have the truth of a moment in your hand, then it slips through your fingers and is lost. When we look into a mirror we think the image that confronts us is accurate. But move a millimetre and the image changes. We are actually looking at a never-ending range of reflections. But sometimes a writer has to smash the mirror — for it is on the other side of that mirror that the truth stares at us. I believe that despite the enormous odds which exist, unflinching, unswerving, fierce intellectual determination, as citizens, to define the *real* truth of our lives and our societies is a crucial obligation which devolves upon us all. It is in fact mandatory. If such a determination is not embodied in our political vision we have no hope of restoring what is so nearly lost to us — the dignity of man."

<u>Pinter</u> clearly states that truth is elusive in drama: "a writer has to smash the mirror" for the truth to stare at us.

But when it comes to the society, in which we participate as citizens, we ought to stand firm in defining *real* truth, which is highly mistreated by politicians.

PART 5

DEFINITION OF SCIENCE

"Science is any system of knowledge that is concerned with the physical world and its phenomena and that entails unbiased observations and systematic experimentation. In general, science invokes a pursuit of knowledge covering general truths or the operation of fundamental laws" (*Britannica Academic*, Encyclopaedia Britannica 2009, 2020).

To be noted is that *truth* is included in the definition of science.

A CHEMIST'S REFLECTIONS IN THE 20th CENTURY

Linus Pauling (1901-1994) has been listed as one of "The twenty greatest scientists of all time." His main contribution was the use of physics to explain chemistry, which revolutionized the latter. He also made seminal contributions to crystallography, nuclear physics and immunology, being considered one of the founders of molecular biology.

Pauling is one of the few scientists who received two Nobel Prizes (Chemistry 1954; Peace, 1962).

One of Pauling's last interviews was given in April 1994 (four months before he died) to several colleagues. George B. Kauffman and Laurie M. Kauffman (1994) report that this was an encounter with a man of "boundless wide-ranging curiosity, self-confidence, outspoken iconoclasm, humor, fearlessness... and an abiding belief in the application of reason to scientific and human affairs."

Some excerpts in condensed form:

If "someone has made an observation that does not fit into my picture of the universe" "I find some way of fitting it in."

"Individual scientists who are successful in their work are successful for different reasons."

"I am especially interested in research that involves one person solving one problem."

"Nowadays we have big physics — million — dollar... or billion — dollar physics. Papers are published in 'Physical Review Letters' with more than a hundred authors... but there is still a good number of people, theoretical physicists and chemists, who continue to work the old way of the individual trying to have an idea that will lead to the solution of some problem."

What transpires from this interview and from previous statements, in connection with his work, is: a boundless curiosity, an enormous self-confidence, great courage, and a need to destroy cherished ideas in science and society.

PAVLOV'S TESTAMENT — SCIENCE DEMANDS A WHOLE LIFE

The British physicist J.D. Bernal, who was also a crystallographer, and one of the pioneers of molecular biology, wrote "The Social Function of Science" in 1939. This book, which appeared at the beginning of World War II, had a great impact on the reorganization of British science after the War. Its influence extended to the Scandinavian countries which also adopted his proposed reforms.

Bernal included in his book the description of what is science by Ivan Pavlov, the Russian physiologist (1849-1936), who received the Nobel Prize in Medicine in 1904 and who had died recently. Pavlov's words were extracted from "Pavlov and his School" by Professor Y.P. Frolov. M.D. Bernal introduced them with the following comment: "The needs and the possibilities of Soviet science are movingly expressed in Pavlov's last testament to his students":

"What is it that I would wish the young men and women of my country who have dedicated themselves to science? first of all — consistency. Of this very important condition of fruitful scientific work I can never speak without emotion. Consistency, consistency and again consistency. From the very beginning of your work, train yourself to strict consistency in the acquirement of knowledge. Learn the ABC of science before you attempt to scale its peaks. Never embark on what comes after without having mastered what goes before. Never try to cover up the gaps in your knowledge, even by the boldest guesses and hypotheses. Such a bubble may delight your eye by its play of colors, but it will inevitably burst and you will be left with nothing but confusion. Train yourself to reserve and patience. Learn to do the heavy work that science involves. Study, compare, accumulate facts. Be the wing of a bird never so perfect, it would never bear her aloft without the support of the air. Facts are the scientist's air, without which he would never be able to fly. Without facts, your theories are labour in vain. But in studying, experimenting and observing — try not to remain at the surface of the facts. Do not turn yourself into a museum custodian of facts. Try to penetrate into the secret of their origin. Steadfastly seek the laws that govern them. The second thing is — modesty. Never think that you already know everything. And however high the esteem in which you are held, always have the courage to say to yourself: 'I am ignorant.' Do not allow pride to take possession of you. It will cause you to be obstinate when you should be conciliatory. It will cause you to reject useful advice and friendly help. It will prevent you from taking an objective view. In the collective which I have to guide, everything depends on the atmosphere. We are all harnessed to a common cause and each of us helps it forward to the extent of his strength and possibility. With us it is often impossible to distinguish what is 'mine' and what is 'yours.' But our common cause only gains thereby. The third thing is — passion. Remember that science demands a man's whole life. And even if you had two lives, it would not be enough. Science demands from man great intensity and deep passion. Be passionate in your work and searchings."

This is a unique document that describes what is science with the utmost clarity. The emphasis is on several fundamental aspects: 1) Consistency, 2) learn the ABC of science before you attempt to scale its peaks, 3) reserve and patience, 4) seek the laws that govern facts, 5) modesty, 6) do not allow pride to take possession of you, 7) deep passion, 8) science demands a man's whole life.

Few scientists have expressed it so well.

THE ANTITHETIC NATURE OF A SCIENTIST'S QUALITIES AND FAULTS

It is not easy to understand or to explain the processes that guide a simple and common phenomenon, *viz*: how many competent students and young research workers are constantly eliminated from the scientific careers that they had entered full of expectations or in which they had already achieved some status. Most of us think that we know how science is made. However, a closer analysis discloses the complexity of the processes involved in the pursuit of original research and exposes the difficulties in disentangling the several factors that participate in its making.

Most students who enter the University have a series of qualifications in common. They are: well educated, normally intelligent, partly dedicated and with a certain degree of self-esteem. Such average values lead them to think that they may succeed in the profession that they have chosen. However, after a few years, the situation changes and they are confronted with another intellectual scenario, that is quite different from their original dream. They may even finish having to labor in an unrelated profession.

What is tragic, is that they notice that other colleagues, who were less intelligent, less determined, and many times less educated, went forward in their scientific careers whereas they themselves were rapidly set aside. As a result many people maintain a bitter attitude towards life and chide the University which became a disappointment to them. Still more intriguing was the situation in which a student had gone as far as to get a doctor's degree and had published a series of papers, yet that seemed to be not enough. One was removed, abruptly from the scientific community, to work in industry, to become a teacher, a priest or even discarded, to join the group of the unemployed. What had really happened? What had been missed or what had never occurred?

Already as children go through the first years of school their parents are impressed by their performance. It is natural that they are proud of their achievements and envision a bright future for them. If any of the children draws well he will be expected to be a new Picasso, if another plays the piano at a tender age it is expected that he will be a new Mozart, and if yet another is good in chemistry she is seen as the embryonic form of a Marie Curie.

Here, lies the primary germ of the confusion. This attitude towards children reveals a lack of the understanding of the mechanisms that lead to achievement in a specific area of human endeavor. Parents start by assuming erroneously, that the endowment of a single positive ability is enough to master a subject in its totality.

The possession of a given quality, the strongest or the best it may be, leads no one anywhere. A series of other qualities must accompany the main one. Will it be enough to add two extra ones, such as intelligence or imagination? Not even these seem to do. Let us then add a strong memory. Not even such a person would be equipped to achieve a brilliant career. Unfortunately, that does not seem to be enough. What comes as a revelation is that the possession of negative properties is equally important and indispensable to achieve an outstanding position.

To put it short, at least more than a dozen qualities and more than a dozen faults seem to be necessary to perform original and significant work. Perhaps, most surprising is that these qualities and faults are antithetical, opposing each other at every moment in our lives. One may need to use a positive one at a specific moment and within five minutes a negative one. A permanent switch between alternatives seems to be necessary.

IT SEEMS THAT SCIENTISTS ARE LIKE EVERYBODY ELSE

With so many different faults and qualities, it seems that in the end, scientists are not very different from everybody else. This is undoubtedly true. The differences between all of us are only hair thin. It is only minor variations in the details of the personality that make the whole difference.

The major difference between the scientist, and what could be called the general citizen, lies in the uniqueness of the constellation of qualities and faults that happen to have combined in a single person. If one is missing, or one has been added, the result may be totally different. This is why outstanding scientists are such rare individuals and can never be repeated. Moreover, the qualities or faults that one possesses may become more evident in certain circumstances.

This is why most scientists appear to the general public as persons who easily fall from their pedestals and show the ugly face of their human limitations. Outside their own fields of research, scientists are vulnerable and many times are as incompetent as anyone else. Moreover, within their own specialty they are not as objecttive as we may tend to think, their view is strongly tinged by their emotions, political views and cultural background.

The Scottish historian Thomas Carlyle (1795-1881) expressed the same opinion: "I confess, I have no notion of a truly great man that could not be all sorts of men" (Carlyle 1995). But René Descartes (1596-1650), the leading mathematician and philosopher, put it in a clear way: "The greatest minds are capable of the greatest vices as well as of the greatest virtues" (Descartes 1997).

OCCUPATION AND PREOCCUPATION

Just as there are musicians and painters of all categories, there are also scientists of different degrees of competence. Most of them are better called research workers. For the great majority science is a profession like any other, and their position is better defined as an occupation. They could easily have occupied other professions that they would have carried out with equal efficiency and competence.

Yet, among this army of research workers, one finds a very small group, for which science is not an occupation but has been a life's preoccupation. They have been from an early age obsessed by the meaning of the laws that govern natural phenomena, be they in physics, chemistry or biology. They have not been looking for facile solutions or rapid results. On the contrary they have assembled, many times, for over 30 years, the data and the knowledge necessary to bring coherence to an early idea, that was only partly unveiled.

They have gone, through the many distractions and interferences of every day life, only seeing in front of them a given phenomenon like sleep-walkers. They have centered all their energy into elucidating a certain type of scientific information, by means of successive experiments allied to a permanent intellectual probing, that has lasted many times a whole life span.

These are the obsessed. It is only by their enormous canalization of action combined with a permanent spirit of enquiry that they were allowed to arrive at a partly coherent picture of a phenomenon. Victor Hugo described them in a poem when he wrote that they combined a "grand amour" with a "grand labeur" (a great love with a great labor).

WRITING ON HORSEBACK, IN TRAINS OR IN AIRPLANES

Some scientists considered the airplane an ideal place to work. Telephones are not ringing, inquisitive post-doctors are not in sight, secretaries do not interrupt. One cannot ask for a better Eden to write in peace. In addition, a pleasant stewardess serves a gourmet meal accompanied by a carefully selected wine. Out of the window run white clouds that build huge castles in the sky. On one occasion, one discerns, far away in the horizon, the sun rising over the peaks of the Peruvian Andes. On another occasion, the conical top of Mount Fuji builds a silhouette against the sun setting on the Sea of Japan. What better place can one demand to put the final touches on a scientific report or to add some paragraphs to a book that is in progress?

Not only planes are suitable, trains are also "niches," which afford such an isolation from the everyday routine of the laboratory. Sitting comfortably by the window one may write in a few hours journey, what could not be accomplished in days at one's desk.

But even the horseback has its great advantages. Instead of trotting and employing time on trivial discussion about the weather, with a journey's companion, one can use this means of locomotion to write the pages of the "Praise of Folly" (1511, 1993). <u>Erasmus</u> of Rotterdam, tells us, how he, in his journey from Italy to England, wrote his notes on horseback that later became his best book. He describes this event in a letter addressed to his friend <u>Thomas Moore</u>.

Others long before him had used similar solutions. The Italian poet <u>Francesco Petrarca</u> (1304-1374) wrote his discourses when on a sailing boat.

Several authors have described how they have used short intervals, in their everyday life to write each time a small portion of a work, which after years of toil has crystallized into a large volume.

A CLASSIC WORK WAS WRITTEN IN A WEEK

Most creative minds are usually known to posterity by one or two pieces of their work, although their output was in most cases enormous. Erasmus is also an example of such a situation; his writings occupy 10 volumes, one of them containing a collection of his over 3,000 letters. Yet today he is mainly known for his satire of the church, the "Praise of Folly." What a healthy reading for any intellect! It is so clear in its expression, so perfectly formulated, so violent in its exposure of the infamous in society, and at the same time so measured in its understanding of the folly caused by mankind's limitations.

This work was written in a week, as stated in a letter that he wrote to the "distinguished theologian Maarten van Dorp" in 1515 as a reply to Dorp's criticism of his work (Erasmus 1515, 1993). He adds that he wrote the "Folly" as a distraction and that a week was "too long time," for he considered it of minor importance when compared to all his writings. A work, like the Folly, which contains such a concentrated distillation (into 134 pages) of human wisdom, could only be written in a week, if it was already well assembled in Erasmus' mind.

THE FERMENTATION OF IDEAS IN SCIENCE TAKES MANY YEARS

Nicolaus Copernicus (1473-1543) was the Polish astronomer who proposed heliocentric cosmology according to which it was not the Earth but the Sun, which was the center of the planetary system. The geocentric model of Ptolemy had been adopted by the Church and was generally accepted for over 1000 years. His astronomic observations led him to the new idea, as early as 1514, but it took him 30 years to give it final form in the book with the title: "De Revolutionibus," which appeared in 1543. Copernicus' ideas were criticized by other colleagues such as T. Brahe and also by the Church which officially banned his book in 1616 and which until 1835 remained on the list of forbidden books.

Galileo Galilei (1564-1642), the Italian physicist and astronomer demonstrated the correctness of Copernicus' system. He made a telescope that allowed him to observe for the first time in 1610 the mountains of the Moon, the existence of four satellites circling around Jupiter, the phases of Venus and the composite structure of Saturnus. His "Dialogue on the two Chief World Systems — Ptolemaic and Copernican" was not published until 1632. As much as 22 years of permanent concentration on the same subject were necessary (Ian et al. 2002).

In chemistry, Dmitri Mendeleev (1834-1907) had to submerge his mind for 8 years (1861-1869) before he published his chemical tables. His work is one of the best examples of a permanent focalization on a basic idea. At that time, only 60 chemical elements were known. He started in by writing the properties of each one on as separate card. Then, he discovered that there was an indication of order in the way the cards could be assembled, but this order was elusive. When the chemical elements were placed following their relative atomic mass, elements with similar properties, tended to build vertical groups. There was a periodicity. But there were holes in the tables, there was no perfect fitting. Two features were evident. 1) The atomic masses of some elements had to be in error. 2) There were missing elements that had to be discovered if the periodicity really existed. He then predicted the existence of several elements, that would fill the blank spaces. These were soon discovered by other chemists in 1886.

For Charles Darwin (1809-1882) the central idea that he exposed in the "Origin of Species" took 20 years to mature. His voyage to South America on board the ship "Beagle" began in 1831. It was his confrontation with the particular features of the plants and animals on this continent that aroused in him the need to consider a different explanation of evolution than that proposed by Lamarck in 1809. The "Journal" which described his voyage and his observations was published in 1839, but the "Origin of Species," where he elaborated in detail his novel views, only appeared in 1859.

IN ART AN IDEA ALSO TAKES A LONG TIME TO MATURE

In painting there is an iconic case.

Katsuhika Hokusai (1760-1849) was the Japanese painter, considered to be the master of wood block printing. He painted the same motif many times in an effort to express the essence of what he could extract from nature. The largest volcano of Japan, Mount Fuji, which was still in eruption during Hokusai's life, has a conical shape that can be seen against a pure sky, as it rises directly from the sea to a height of over 3,000 meters. Its perfect symmetry, with its top covered with snow in winter, makes it a unique sight, full of beauty and simplicity. The painter was fascinated by this mountain and made a series of paintings that he called "The 36 views of Mount Fuji." Each one conveys a different message that is furnished by the same object. The time of the day, the time of the year, the type of weather, the type of human behavior connected with the mountain, all represent a tremendous effort to find out what a simple natural scene may tell, the receptive mind, the message furnished by natural phenomena.

To understand what he saw in the landscape of Japan, took him a whole life. As one of his biographers points out (Forrer 2004): "In fact, all the works upon which his fame rests were made during the short period between 1830 and 1836, that is between the ages of seventy-one and seventy-seven. That he too considered this period in his life as the true beginning of his career, can be inferred from the brief autobiography he wrote aged seventy-five, which begins: "From the age of six I was in the habit of drawing all kinds of things. Although I had produced numerous designs by my fiftieth year, none of my works done before my seventieth is really worth counting." And he adds "At the age of seventy-three I have come to understand the true form of animals, insects and fish and the nature of plants and trees." These quotations are taken from Hokusai's autobiography, written in 1835, at the age of 75 years (Kerrigan 2016).

MONUMENTAL WORKS

Many outstanding scientists have, in modern times, published over 200 scientific works. These are not newspaper articles, but represent reports on scientific experiments that were the product of a huge labor at the laboratory bench.

However, there are some contributions that even exceed these.

Linus Pauling (1901-1994) was the author of over 1000 publications in the field of chemistry (Kauffman and Kauffman 1994). Many of them revolutionized the knowledge of this science. Another extremely prolific scientist was Ernst Haeckel (1834-1919). He wrote several large books in biology but his astounding contribution is the illustration of his works. His "Challenger Monograph" contains "thousands and thousands" of figures in which are described and portrayed, for the first time, 3,508 new species of radiolarians discovered by him on the Challenger Expedition in 1887.

This intellectual capacity goes back to

Antiquity. Only a small proportion of the works of Roman and Greek literature and science has survived. Wars, fires, earthquakes, church orthodoxy and savage ignorance, led to the destruction of this precious heritage. Titus Livius (59 B.C. - A.D. 17) published not less than 142 books of which only 35 have reached our time.

The founder of modern biography, and one of the most important writers of Antiquity, was Mestrius Plutarchus (b. before A.D. 50 - d. after A.D. 120). We possess today only half of what he wrote but it still fills a dozen volumes. Besides his unique "Parallel Lives" of the leading figures of his time, there are over 70 surviving essays and dialogues that were collected under the name of "Moralia." These represent a precious analysis of ethical values. They are as actual today, as they were 2000 years ago.

The preservation of the work of these authors is mainly due to Arab and Byzantine scholars, who in Baghdad and Constantinople translated the few copies that had survived in the cultural centers of Iraq, Syria and Saudi Arabia during the Middle Ages.

Equally monumental were the works produced by Chinese scholars. Sima Quian (145 - c. 85 B.C.) wrote the "Historical Records" of China in 130 chapters which cover institutions, biographies, political events and history. Another author Du You in 801 A.D. submitted to the emperor an enormous history of Chinese institutions in 200 chapters consisting of over 5,000 pages. During the Ming dynasty, an Encyclopaedia in 11,095 volumes was compiled between 1403 and 1407 (Sullivan 1973).

In painting there was also an artist with a gigantic production. The Spaniard Pablo Picasso (1881-1973) was fully active until his last days, being over 90 years old. He produced on the average three art works a day during a period of 70 years. His painting has been divided into several periods (such as blue period, cubist period) but also he made a great contribution in lithography, ceramics and sculpture. His total work is more vast than that of his contemporary colleagues taken together (20,000 paintings, drawings and sculptures). "The 34 volume catalogue of Picasso's work, compiled by Christian Zervos, and published between 1942 and 1978, while far from complete, reproduces over 16,000 works. Other catalogues document over 600 sculptures and 200 ceramic works; they too are incomplete" (Warncke 2007).

In music the output may have been small in some cases but in others it attained the same extension as in science. Giovanni da Palestrina (1525/6-1594) was an extensive writer of Renaissance music, his complete works cover 33 volumes. Leo Tolstoy (1828-1910) took years to write his main novel "War and Peace." His wife functioned as his secretary and she copied the revisions that he wrote. He went again and again through the original manuscript making additions and corrections, only being satisfied when he reached the 10th and final version.

In the area of painting the situation repeats itself. Henri Matisse, the French painter (1869-1954), wrote in 1949 five years before his death: "It has taken me all this time to reach the stage where I can say what I want to say."

The enormous effort and concentration needed, was crystallized in another formula by the Italian composer Giacomo Puccini (1858-1924): "Art, like life, is a strife, and the one who gives most blood is the one who collects most wisdom." This statement is confirmed by another composer. It took Vaughan Williams (1872-1958), 30 years to compose the opera "The Pilgrim's Progress" (1951), being finished just seven years before his death.

Some writers had their books in their minds, for years, in nearly ready form. They needed only a suitable opportunity to put them on paper. The French writer J.-A. Brillat-Savarin (1755-1826) compiled his views on the science and the art of eating in "The Physiology of Taste." In his classical work, he tells: "When I began to write, my list of contents was already drawn up, and my book complete in my head."

EPILOGUE

GEOMETRY AND THE ORDER OF NATURE WERE THE MAIN PREOCCUPATION OF ARCHIMEDES

It may be appropriate to finish this work by listening to three personalities. They are most different in time and space but they convey to us precious messages based on reason. They are <u>Archimedes</u>, <u>Marcus Aurelius</u> and <u>Rousseau</u>.

<u>Archimedes</u> of Syracuse (c. 287-212 B.C.) was a Sicilian Greek mathematician and physicist and a pioneer of statics and hydrostatics. One of <u>Archimedes</u>' principles states that when a body is wholly or partially immersed in a fluid, it experiences a buoyant force equal to the weight of the fluid displaced. This experimental procedure allowed, for the first time to easily distinguish between objects made of metals having different densities such as gold compared with alloys that imitate it. Best known is his water-screw, which allowed to move water up a slope *i.e.* against gravity. He also used geometrical meth-

ods to calculate the volume of solids. Archimedes is generally considered the finest scientist of the ancient world.

His testament reveals that the order of nature was at the basis of his thought.

It is Plutarch (c. A.D. 50 - c. 120) who in one of his classical biographies "Makers of Rome," reveals to us this precious information on Archimedes' last wish.

"Archimedes asked his friends and relatives to place on his tomb after his death nothing more than the shape of a cylinder enclosing a sphere, with an inscription explaining the ratio by which the containing solid exceeds the contained."

Geometry is the ultimate form of order that pervades our mind, as well as the universe, and which is difficult to demonstrate because order is everywhere but is not total (Lima-de-Faria 2016).

"THERE IS AN INNATE PRINCIPLE OF JUSTICE AND VIRTUE"

One of the great figures of Western civilization was Jean-Jacques Rousseau (1712 -1778) (Dent 1992). For the German philosopher Arthur Schopenhauer "He was the profound judge of the human heart, who drew his wisdom not from books but from life." "He alone was endowed by nature with the gift of being able to moralize without being tedious, for he hit upon the truth and touched the heart."

However, his vast work created enemies everywhere. His book "The Social Contract" was burned in Geneva, and in 1762 another one: "Émile" was condemned by the Faculty of Theology at the Sorbonne and its copies were also burned. Among his enemies was Voltaire who was scornful of Rousseau's views.

The "Confessions" have a history of their own. In late 1761 Rousseau's usual publisher M.-M. Rey urged him to write his autobiography. At first he did not like the idea. Finally he started writing but abandoned the manuscript for long periods. Suddenly his friends became also his enemies. They banned the first two parts because they feared they would be harmed by his writings. One of them was Madame d'Epinay, who suspected that she would not be portrayed to her advantage in them. The result was that the first complete edition of the "Confessions" did not appear until 1789 (11 years after his death). The book opens with a defiant declaration that shows his deep honesty and extraordinary clarity:

"I have resolved on an enterprise which has no precedent, and which once complete, will have no imitator. My purpose is to display to my kind a portrait in every way true to nature, and the man I shall portray is myself... let the numberless legion of my fellow men gather round me, and hear my confessions. Let them groan at my depravities, and blush for my misdeeds. But let each one of them reveal his heart at the foot of Thy throne with equal sincerity, and may any man who dares, say 'I was a better man than he'."

In his book "Émile" he adds some fundamental statements concerning conscience: "There is in the depths of souls... an innate principle of justice and virtue." There is in all of us the capacity to admire and to rejoice in acts of justice, of kindness and of moral beauty. Dent (1992) adds that Rousseau's "voice" of conscience, is an expression, an awareness, of the demands of this inner order of man.

THE EMPEROR MARCUS AURELIUS HAS THE LAST WORD

Emperors are usually common people of little distinction. Charlemagne, the emperor of France, Germany and part of Italy (742-814 A.D.) was an analphabet. He could hardly write his name on the document when he came to Rome to be crowned emperor by the Pope (Padamsee 2002).

This was in the Middle Ages, during the Roman Empire, the city had not less than 10 public libraries and the Greek and Roman authors were, not only a source of knowledge, but of permanent debate. Besides, Roman intellectuals wrote works that remain a permanent source of reference: Plutarch, Cicero, Caesar, Horace and many others. Among them, one stands out as unique, the Emperor Marcus Aurelius (121-180 A.D.) who wrote his "Meditations."

He was not interested in power. As he became emperor he appointed Lucius Verus as his colleague in government.

There were thus, two emperors ruling for the first time in Roman history. Natural disasters: famine, plagues and floods, followed in succession, accompanied by invasions of barbarians. Verus died soon and Marcus Aurelius had to join his legions on the Danube River alone. It is then that he consoled himself by writing a series of reflections that he called simply "To Himself."

They "reveal a mind of great humanity and natural humility" (Staniforth 1977). He has also been described as "perhaps the most beautiful figure in history" (Hammond 2006). His chief aim and end was the attainment of personal virtue.

Marcus Aurelius never went to school, his highly educated family made sure that he had excellent tutors.

Let us listen to Marcus Aurelius, who considered the Roman gods as a doubtful source, searching instead for reason: "Have you reason? 'I have.' Then why not use it? If reason does its part, what more would you ask?" "Anything that distracts you from the fidelity to the Ruler within you — means a loss of opportunity for some other task." This is "the deity that dwells within you."

"If you do the task before you, always adhering to strict reason, with zeal and energy and yet with humanity, disregarding all lesser ends and keeping the divinity within you pure and upright... from this course no man has the power to hold you back."

"What does not corrupt a man himself cannot corrupt his life."

"Never become unduly absorbed in things that are not of the first importance."

"Evil comes not from the mind of another... but from the part of yourself which acts as your accessor of what is evil."

"The vulgar confine their admiration chiefly to things of an elementary order."

The list could be made longer. What is significant is that the Meditations were obligatory reading for generations especially during the 1700s and 1800s. At present most people never even have heard of them. However, the highly educated continue to consider this work a precious source. In China, Premier Minister Wen Ja Bao told, in an interview, that when he traveled, the Meditations of Marcus Aurelius was his best companion.

REFERENCES

"art" Britannica Academic, Encyclopaedia Britannica, 21 Dec. 2016. academic.eb.com/levels/collegiate/article/art/389134. Accessed 28 May. 2020.

"music" Britannica Academic, Encyclopaedia Britannica, 9 Aug. 2018. academic.eb.com/levels/collegiate/article/music/110117. Accessed 28 May. 2020.

"science" Britannica Academic, Encyclopaedia Britannica, 24 Jul. 2019. academic.eb.com/levels/collegiate/science/66286. Accessed 28 May. 2020.

"truth" Britannica Academic, Encyclopaedia Britannica, 20 Apr. 2009. academic.eb.com/levels/collegiate/article/truth/473522. Accessed 28 May. 2020.

Abraham, G. 1943. Achille-Claude Debussy. In: Lives of the Great Composers, vol. III. Penguin Books, New York, USA.

Bernal, J.D. 1939. The Social Function of Science. G. Routledge, London, UK.

Calvocoressi, M.D. 1943. Modeste P. Moussorsky. In: Lives of the Great Composers, vol. II. Ed. A.L. Bacharach. Penguin Books, New York, USA.

Carlyle, T. 1995. On Great Men. Penguin Books.

Chilvers, I. 2003. Editor. Oxford Concise Dictionary of Art and Artists. Oxford University Press, Oxford, UK.

Chipp, H.B. 1968. Theories of Modern Art. University of California Press, Berkeley, USA.

Cynx, J. 2004. Are Songbirds Pythagoreans? In: Nature's Music. The Science of Birdsong. Eds. P. Marler and H. Slabbekoorn. Elsevier, Amsterdam, Holland. Page 218.

Dent, N.J.H. 1992. A Rousseau Dictionary. Blackwell, Oxford, UK.

Descartes, R. 1997. Discourse of the Method. Descartes, Key Philosophical Writings. Wordsworth Classics of World Literature, Ware, UK.

Dunn, J. and Clark, M.A. 1999. Life music: the sonification of proteins. Leonardo, vol. 32(1): 25-32.

Erasmus of Rotterdam (1511). 1993. Praise of Folly. Penguin Books, London, UK.

Erasmus of Rotterdam (1515). 1993. Letter to Maarten van Dorp. Penguin Books, London, UK.

Evans, E. 1943. Louis Hector Berlioz. In: Lives of the Great Composers, vol. II. Penguin Books, New York, USA.

Ferry, G. 2007. Max Perutz. Cold Spring Harbor Laboratory Press, Cold Spring Harbor, New York, USA.

Flam, J.D. 1973. Matisse on Art. Phaidon, London, UK.

Forrer, M. 1988. Hokusai. Prestel, London, UK.

Gahr, M. et al. 1993. Estrogen receptors in the avian brain: survey reveals general distribution and forebrain areas unique to songbirds. J. Comp. Neurol. 327: 112-122.

Ghiselin, B. 1952. The Creative Process. A Mentor Book, New American Library, New York, USA.

Gibson, M. 2003. Good Vibrations: The Sounds of Abstraction. International Herald Tribune, November 15-16, 2003. New York, USA.

Goethe, J.W. von 1998. Maxims and Reflections. Penguin Books, London, UK.

Greenwood, E.B. 1975. Tolstoy: The Comprehensive Vision. Dent and Sons, London, UK.

Gribbin, J. and Gribbin, M. 2018. Richard Feynman, A Life in Science. Icon Books Ltd, London, UK.

Grout, D.J. and Palisca, C.V. 2001. A History of Western Music. W.W. Norton and Company, New York, USA.

Haeckel, E. 1904. Kunstformen der Natur. Vienna, Austria. Art Forms in Nature. 1974, Dover Publications, New York, USA.

Haesler, S. et al. 2007. Incomplete and inaccurate vocal imitation after knockdown of *FoxP2* in songbird basal ganglia nucleus area X. PLoS Biol. 5: 2885-2897.

Hammond, M. 2006. Marcus Aurelius Meditations. Penguin Classics, London, UK.

Hobsbawm, E. 1998.The Age of Capital (1848-1875). Abacus, London, UK.

Hornblower, S. and Spawforth, A. 1999. The Oxford Classical Dictionary. Oxford University Press, Oxford, UK.

Honour, H. and Fleming, J. 2002. A World History of Art. 6th Edition. Laurence King Publishing, Laurence, USA.

Hughes, H. 1943. Frederic F. Chopin. In: Lives of the Composers, vol. II. Penguin Books, New York, USA.

Ian, D. et al. 2002. The Cambridge Dictionary of Scientists. Cambridge University Press, Cambridge, UK.

International Chicken Genome Sequencing Consortium (incl. L. Andersson and H. Ellegren) 2004. Sequence and comparative analysis of the chicken genome provide unique perspectives on vertebrate evolution. Nature 432: 695-715.

Jean, R.V. and Barabé, D. 1998. (Eds.) Symmetry in Plants. World Scientific, Singapore, London.

Katz, V.J. 1993. A History of Mathematics. Harper Collins, College Publishers, New York, USA.

Kaufmann, H. 1947. The Little Guide to Music Appreciation. Grosset and Dunlap Publishers, New York, USA.

Kauffman, G.B. and Kauffman, L.M. 1994. Linus Pauling: Reflections. American Scientist 82:522.

Kerrigan, M. 2016. Visions of Fuji. Flame Tree Publishing, London, UK.

Kropotkin, P. 1962. En Anarkists Minnen. Forum, Stockholm, Sweden.

Lima-de-Faria, A. 2012. Molecular Geometry of Body Pattern in Birds. Springer, Heidelberg, Germany, New York, USA.

Lima-de-Faria, A. 2016. Order is present at every level but is not total. Theoretical Biology Forum 109(1-2). Pisa, Rome.

Macdonald, D. 2002. The New Encyclopedia of Mammals. Oxford University Press, Oxford, UK.

Marler, P. 2004. Science and birdsong. In: Nature's Music. The Science of Birdsong. Eds. P. Marler and H. Slabbekoorn. Elsevier, Amsterdam, Holland.

Matisse, H. 1954. Looking at life with the eyes of a child. Art News and Review, London, UK.

Mondrian, P. 1919. Natural Reality and Abstract Reality. *De Stijl*, Amsterdam, The Netherlands.

Myers, R.H. 1943. Peter I. Tchaikovsky. In: Lives of the Great Composers, vol. III. Penguin Books, New York, USA.

Ohno, S. and Jabara, M. 1985. Repeats of base oligomers *($N = 3n +- 1$ or 2)* as immortal coding sequences of the Primeval World: Construction of coding sequences is based upon the principle of musical composition. Chemica Scripta 26B: 43-49.

Ohno, S. and Ohno, M. 1986. The all pervasive principle of repetitious recurrence governs not only coding sequence construction but also human endeavor in musical composition. Immunogenetics 24: 71-78.

Ohno, S. 1987. Repetition as the essence of life on this earth: music and genes. In: Haematology and Blood Transfusion. Vol. 31. Modern Trends in Human Leukemia VII. Springer-Verlag, Berlin, Heidelberg, Germany.

Ohno, S. 1988. Of words, genes and music. In: NATO ASI Series, Vol H23. The Semiotics of Cellular Communication in the Immune System. Springer-Verlag, Heidelberg, Germany.

Ohno, S. 1988. On periodicities governing the construction of genes and proteins. Animal Genetics 19: 305-316.

Padamsee, H.S. 2002. Unifying the Universe. Institute of Physics Publishing, Bristol, USA.

Perutz, M.F. 2003. I Wish I'd Made you Angry Earlier. Essays on Science, Scientists, and Humanity. Cold Spring Harbor Laboratory Press, Cold Spring Harbor, New York, USA.

Pinter, H. 2005. Art, Truth and Politics. Nobel Lecture. The Nobel Foundation, Stockholm, Sweden.

Plutarch 1965. Makers of Rome. Penguin Classics, London, UK.

Poincaré, H. 1906 (1943). La Science et l'Hypothèse. Flammarion, Paris. France.

Poincaré, H. 1908 (1947). Science et Méthode. Flammarion, Paris, France.

Read, H. 1950. The Meaning of Art. Penguin Books, London, UK.

Rodin, A. 1946. L'Art. Entretiens Réunis par Paul Gsell. Mermod, Paris, France.

Rolland, R. 1928. Vie de Tolstoy. Librairie Hachette, Paris, France.

Sadie, S. 2014. New and Expanded Classical Music Encyclopedia. Flame Tree Publishing, London, UK.

Salter, L. 1978. The Gramophone Guide to Classical Composers and Recordings. Salamander Books, London, UK.

Scholes, P.A. 1977. The Concise Oxford Dictionary of Music. Oxford University Press, Oxford, UK.

Schopenhauer, A. 1970. Essays and Aphorisms. Penguin Books, London, UK.

Shen, P. et al. 1995. An atlas of aromatase mRNA expression in the zebra finch brain. J. Comp. Neurol. 360: 172-184.

Staniforth, M. 1977. Marcus Aurelius Meditations. Penguin Books, London, UK.

Stanley, J. 1995. Bonniers Bok om Klassisk Musik. Bonnier Alba, Stockholm, Sweden.

Stravinsky, I. 1936. An Autobiography. Simon and Schuster, New York, USA.

Sullivan, M. 1973. The Arts of China. Cardinal, London, UK.

Thomson, J.A. 1920. The System of Animate Nature. Williams and Norgate, London, UK.

Toye, F. 1943. Giuseppe Verdi. In: Lives of the Great Composers (vol. 3). Penguin Books, New York, USA.

Waetzoldt, W. 1938. Tu y el Arte. Spanish translation, Labor, Barcelona, Spain.

Warncke, C.-P. 2007. Pablo Picasso 1881-1973. Taschen, Köln, Germany.

Weinberg, S. 2015. To Explain the World. Penguin, Random House, London, UK.

West, M.J. and King, A.P. 1998. Mozart's starling. American Scientist 1998: 81-89. Sinauer Associates, USA.